Limited edition
CAPM "EXAM" CERTIFICATION PREP
Go Above and Beyond. Boost Your Value in Personal Development.
START YOUR CAREER FROM NOW

CAPM Exam Certification Prep [Pmbok Guide 2021-22]

The Idiot-Proof Guide that Helped over 1,347 Lazy Students Pass the CAPM Exam on the First Attempt and Start Working with Multi-Million Companies Right Away

The Crazy Driver

CONTENTS

PART I

PART II Managing Cisco Routers

PART III Authentication, Authorization, and Accounting (AAA)

Introduction

What is CAPM?

- CAPM represents Affirmed Partner in undertaking The board, an industry-perceived certificate for passage level task supervisors or venture chief wannabes. This accreditation test is directed by PMI, Task The board Establishment. You need to contemplate PMBOK most recent release (fifth, at the present time) to plan for this test.

  - **Which implies that you can utilize any PMP arrangement asset for showing up for CAPM test too?**

- PMI's affirmation and certifications are set up by specialists so they are useful and significant. These are recognized by their worldwide turn of events and application, which implies that anybody across the globe can take this accreditation and apply the information to effectively oversee projects in any industry. PMBOK is a structure (and not a technique) with cycles of undertaking the executives cut and diced across Information Territories (Degree, Danger, Correspondence, etc.) and Interaction Gatherings

(Starting, Arranging, etc.) making it pertinent across businesses. Notwithstanding, there might be project the executives devices or practices explicit to an industry that you may have to embrace as relevant.

- There are numerous shared characteristics among CAPM and PMP test, however for simplicity of utilization of data this post subtleties all that you require knowing for CAPM at one spot.

- This book is designed to help you prepare for the Cisco SECUR certification exam. The SECUR exam is the first in a series of five exams required for the Cisco Certified Security Professional (CCSP) certification. This exam focuses on the application of security principles with regard to Cisco IOS routers, switches, and virtual private network (VPN) devices.

- There are two rules you need to meet to show up for the CAPM test: 1. you need to have Auxiliary certificate (secondary school or same) or more 2. You need to have 1500hrs of expert venture the executives experience in a group OR 23 contact long periods of formal task the board instruction

- Any college degree in venture the board discipline or contact hour preparing from a PMI enlisted instruction supplier (REP) can be considered towards 23hrs of formal task the executives training. There are numerous different roads to get this schooling, for example,

- Schooling given by PMI parts
- Manager/organization supported preparing programs
- Courses via Preparing schools
- Online courses, as long as they permit you to take end obviously test and you pass it

- **How to apply for CAPM test?**

- You can either apply on the web or by sending paper based application to PMI. While online applications are handled in 24hrs, you need to hang tight for around 10 business days on the off chance that you send by post.
  - Note that PMI suggests that you apply on the web.
- Note - you need to peruse and consent to PMI code of morals and expert direct (from the handbook accessible on PMI.org) prior to applying for the test.

- **Test Expenses**

- The most ideal route is to take up PMI participation and profit rebate on test expenses. This gets a good deal on acquisition of PMBOK control and some other venture the board books or assets too.

- Expecting you are a part and applying for online assessment you would be paying $225 (185 Euros).

- Download the handbook from PMI webpage to realize charge structure for whole blend of part/non-part, on the web/disconnected, assessment/reconsideration, etc.

- You can pay for test charges utilizing Visa, check, cash request or wire move to PMI.

- Note - PMI can discount your test charge in the event that you demand by composed methods at any rate 30 days preceding the test qualification termination date. You'll lose $100 handling expense in the event that you have not booked for the test as of now.

- **What is this Review cycle?**

- PMI arbitrarily chooses applications for review. In the event that your application is chosen you will get 90 days to submit following data:

1. Duplicates of formal schooling

2. Marks of your venture administrators/managers for the activities you have referenced on application

3. Duplicates of certificate for finishing of 23hr contact instruction

- **Re-certification?**

- Indeed. When you breeze through CAPM test, your affirmation will be legitimate for a time of 5 yrs. toward the finish of this you should re-confirm yourself to keep having a legitimate CAPM capability after your name. This re-confirmation should be finished during fifth year of approval period and not prior.

- **<u>Test Content</u>**

- CAPM test will have 150 inquiries, to be replied in 3 hrs. These inquiries are exclusively founded on most recent PMBOK (fifth version at the present time) control and are checked through psychometric examination.

- Nonetheless, not all inquiry goes towards evaluating your presentation. 15 inquiries out of these 150 will be 'pretest' questions. These are not considered for evaluating on the test and these are test inquiries to be remembered for the future assessment, in view of appraisal of how well they have been perceived and replied by understudies.

- Note – you can take break(s) during tests yet the clock won't be halted. So use breaks admirably, to crush tedium and get spirit once more.

- Here is the dispersion of test inquiries across PMBOK sections.

PMBOK Chapter #	Chapter Name	% Questions
3	Project Management Processes	15%
4	Project Integration Management	12%
5	Project Scope Management	11%
6	Project Time Management	12%
7	Project Cost Management	7%
8	Project Quality Management	6%
9	Project Human Resource Management	8%
10	Project Communications Management	6%
11	Project Risk Management	9%
12	Project Procurement Management	7%
13	Project Stakeholder Management	7%
		100%

CAPM Assessment Arrangements

- When your application is acknowledged, expense is acknowledged and review (if relevant) is sufficiently done, PMI will send you with PMI Qualification ID, assessment legitimacy period (which is one year) start and end dates and planning guidelines by email.

- Tests are conveyed by parametric focuses. You need to plan for the test at www.Prometric.com/pmi dependent on the data you got from PMI and dependent on accessibility of openings at the middle you wish to take the test.

- Upon the arrival of taking test

- You need to convey the accompanying to Parametric focus -

- Government provided character card

- You're most recent photo

- Show up at any rate 30 minutes early at Parametric focus. Sign in, show distinguishing proof and give the special Test ID that PMI sent you by email. You won't be permitted to take any things like number cruncher, food, sweater, books, and sacks and so on inside the test room.

  - The accompanying things will be given by parametric staff –

  - Number crunchers are incorporated into CBT (PC based test) test

  - Scrap paper and pencils

Erasable note sheets and markers

- **Test Results**

- Results are given following consummation of the test. Results are accounted. Pass/bomb score dependent on generally execution

- 3-point capability scale rating for every one of the sections: Capable, Respectably capable, Beneath capable

- Note – in view of the capability scale you can know your solid and feeble territories of PMBOK prospectus.

The chapters of the book cover the following topics:

- Chapter 1, "Organization Security Fundamentals" – Part 1 is an outline of organization security by and large terms. This part characterizes the extent of organization security and talks about the fragile "difficult exercise" needed to guarantee that you satisfy the business need without bargaining the security of the association. Organization security is a persistent cycle that ought to be driven by a predefined hierarchical security strategy.

- Chapters 2, "Assault Dangers Characterized and

Itemized" — Part 2 examines the potential organization weaknesses and assaults that represent a danger to the organization. This section furnishes you with a superior comprehension of the requirement for a successful organization security strategy.

- Chapter 3, "Safeguard Top to bottom" — Up to this point, an organization was viewed as secure on the off chance that it had a solid border protection. Organization assaults are getting considerably more unique and require a security pose that gives guard at numerous levels. Section 3 examines the ideas that coordinate all the security segments into a solitary, extremely successful security procedure.

- Chapter 4, "Essential Switch the executives" — this section subtleties the organization of the Cisco IOS switch and examines the IOS firewall include set. This section centers on the fundamentals undertakings that are needed to deal with an individual Cisco IOS switch.

- Chapter 5, "Secure Switch Organization" — this part discloses how to tie down the authoritative admittance to the Cisco IOS switch. It is essential to tie down this admittance to forestall unapproved changes to the switch.

- Chapter 6, "Validation" — this part examines the wide range of sorts of confirmation and the favorable circumstances and weaknesses of each kind.

- Chapter 7, "Verification, Approval, and Bookkeeping" — AAA has become a vital segment of any security strategy. AAA is utilized to confirm which clients are interfacing with a spe-cific asset, guarantee that they are approved to perform mentioned capacities, and track which activities were performed, by whom, and at what time. Section 7 talks about the combination of AAA administrations into a Cisco IOS climate and what AAA can essentially mean for the security stance of an organization.

- Chapter 8, "Arranging Span and TACACS+ on Cisco IOS Programming" — TACACS+ and Range are two key AAA advancements upheld by Cisco IOS Programming. Section 8 talks about the means for arranging TACACS+ and Range to speak with Cisco IOS switches.

- Chapter 9, "Cisco Secure Access Control Worker" — This part depicts the highlights and building segments of the Cisco Secure Access Control Worker.

- Chapter 10, "Organization of Cisco Secure Access Control Worker" – This section examines the establishment and arrangement of the Cisco Secure Access Control Worker on a Microsoft Windows 2000 Worker.

Figure I-1 *Completing the Chapter Material*

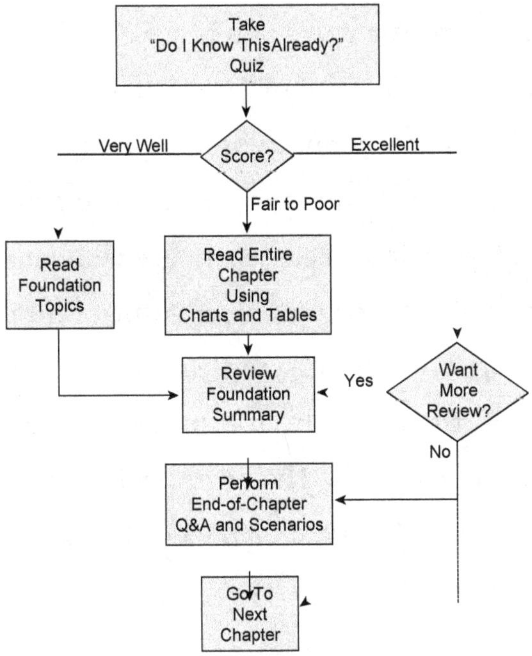

CD-ROM-based practice test, this book incorporates a Compact disc ROM containing a few intuitive practice tests. It is prescribed that you keep on testing your insight and test-taking abilities by utilizing these tests. You will find that your test-taking abilities will improve just by proceeded with openness to the test design. Remember that the likely scope of test questions is boundless. Consequently, your objective ought not to be to "know" each conceivable answer however to have an adequate comprehension of the topic that you can sort out the right answer with the data gave.

Explanation: The fitting reaction is 'stays as an argument against the property'. A repairman's lien is a particularly earth shattering commitment collection mechanical assembly that can be used by project laborers and workers that put in labor or materials to 'improve' authentic property. It is essentially a legitimate document that interfaces with veritable property and gives an ignored laborer for employ a security interest in the property. You ought to understand that a professional's lien reliably runs with the property. If a property is sold, the repairman's lien will not vanish. The argument stays against the property until the lien is totally satisfied. It doesn't have any effect who guarantees the property.

The errand of costs between the buyer and the dealer toward the end is called:

Explanation: The suitable reaction is 'customization'. Customization is the bit of money toward the finish of a property. This tally is consistently done by a land dealer, land salesperson, or land legal counselor. Eventually, this request should fill in as an update that all test-takers need to contribute the energy and effort to learn key land terms. Despite your express, in light of everything, you will run into explicit requests on your licensure test that present for implications of land industry phrasing.

A house sells for $195,000. The outright commission rate is 6%, of which the posting office gets 2.5% and the selling office 3.5%. Susan, the selling trained professional, gets a 60% piece of the commission. What sum will she secure as commission dollars?

Explanation: The proper reaction is '$4,095.00'. The preparation question is a fantastic delineation of such a figuring you might be drawn nearer to wrap up during your state's property allowing test. Here is the way you do it: Start with the total assessment of the arrangement. For the present circumstance, the house sold for $195.000. At that point, you need to know the hard and fast assessment of the commission got by the selling office. For the inspirations driving discovering this arrangement, we were unable to think often less about the posting office's reward. To move towards the last answer, take the certified arrangement regard and increment by percent of commission got by the relevant office ($195,000 copied by 0.035). The commission gained by the selling office is

$6,825. Susan gets 60% of this entirety. All things considered, she gets $6,825 copied by 60% (.60), or $4,095.

One of the activities you can take from this preparation question is that just one out of each odd number associated with the concise will be pertinent to finding the arrangement. In this model, the total commission (6 percent) and the posting office commission (2.5 percent) are essentially extra information; these numbers will not be used. Make an effort not to expect that each number ought to be associated with your calculation to discover to the right arrangement.

This activity manual is unquestionably used

As an improvement to the substance in a fundamental land course, by one who has completed the land course and needs to overview while clutching take the state test, or by one not tried any traditional class anyway who needs to resuscitate principal authentic space considerations and thoughts.

Whatever the clarification Questions and Answers imitates the certifiable appraisal and enables you to advance toward the test with complete trust in your ability to pass.

Not all states use a comparative appraisal. The Relationship of Land Grant Law Specialists (ARELLO) invigorates all land regulatory workplaces to require guaranteed appraisals to test the competency of the up-and-comer. Confirmed test providers fuse AMP, PSI, and Pearson VUE.

THE Point

This book involves distinctive choice inquiries and answers apportioned into subjects

Unsurprising with those subjects covered on most land approving assessments. Each major topic has subchapters and a movement of requests followed by illustrative answers. Additionally, there are addendums that cover standard state express allowing laws, math, review tests, and test last, most significant tests for salespersons and delegates.

Quick Study GUIDE

If you end up with this book nearby and reasonably concise period in which to prepare, or in case you trust you are as of now masterminded, we suggest you first take the review tests and practice last appraisals in the back of the book. If you miss requests on a specific point, center your abundance examination time on those particular zones of the book.

We have furthermore included unequivocal quick study questions (allocated by the disguised box) that pass on the basic concentrations from each part. Complete these smart.

Dealer QUESTIONS

For a quick expert review, certain vendor level requests are alloted with a bull's-eye in the edge. In any case, mediator contenders ought to regardless react to the aggregate of the requests in this book and not simply those set apart with a bull's-eye. Delegate tests cover comparable focuses as those found on an agent's evaluation and consolidate additional requests that oversee office affiliation, office the executives, and human resource headway.

We should turn out a segment of such specialist questions and answers:

In the event that the expenses of an office, involving 24 full-time accomplices, are each year

$480,000, what is the work region cost? The plan is directed by isolating the amount of full-time accomplices (24) into the yearly expenses of $480,000. The proper reaction, by then, is $20,000.

What whole is remaining after all commissions, arrangements, and reference charges are paid out? The suitable reaction is association dollar.

In the event that a specialist expected to keep an essential separation from the possible loss of individual assets from a claim against the lender firm similarly as twofold assessment assortment, the BEST kind of business development would be:

a. Art S association.

b. Restricted association.

c. sole possession

d. general affiliation

The 100% commission blueprint is By and large priceless to whom? The proper reaction is an expert whose productivity is by and large high. Business applications may demand up-and-comers.

A. Age and race.

B. Religion.

C. Previous workplaces.

D. Companion's work.

The proper reaction is C. past workplaces. Moreover, shipper contenders should give interesting thought to Part D (Move of Property Ownership) and Part E (Land Lender).

Directions to Use THIS BOOK

There are two strategies that you can use to enhance your sufficiency and energize study and learning. It is really a matter of individual tendency as for which approach is best for you.

One method looks like eating an elephant: take each eat thusly. Answer only five or then again six requests at the same time, and after each set of requests, investigate the fitting reactions by then do another set, and so forth this system will build your conviction as you come and give your brief info. Now and again, if you answer colossal quantities of requests at the same time without investigating the fitting reactions until later, you can't recall why you made the proper reaction choice you did.

The other strategy is to answer the entire test before looking at the suitable reactions, allowing one second for each question. Various people feel this not simply shapes scrutinizing speed yet moreover allows them to differentiate the rate right and the test objective (typically 70% to 75%) and definitely grows their trust in their testing capacities. Whichever system is used, mark your answers on an alternate piece of paper so when you need to restudy the material, you can't see the previous answers.

In addition, in case you save this paper, you can take a gander at the results the ensuing time around.

Exactly when you have ruled the whole of the parts and supplement material, don't hesitate to complete the preparation last appraisals. Hold adequate ceaseless freedom to take the last and review tests under reenacted testing conditions.

A Strategy FOR Analyzing

Sort out your examination time into two-to-four-hour pieces. Cover no more than four review parts at the same time. Long assessment gatherings are less productive. Sadly, basically scrutinizing the review material will not set you up for the test. To appear at a concentrated cognizance of the material, it is wise to chart your assessment material and make up in your cerebrum subject-related legitimate/sham requests.

Introduction Additionally, in the wake of scrutinizing, use the language words from each review part to make streak cards and subsequently use those cards to test you. In extension to using this book, in case you are in a relicensing class, structure an investigation pack with a couple of others.

The testing providers weight land lender, heavier than by far most of the various areas. Twofold your review time on association, postings, valuation of land, and arrangements. Entertainment Test- Taking Tips

- Find a quiet recognize (no phones or outside interferences) to venture through the tests in this book.

- Never leave a request clear. Restricted your choices and make an educated gather.

- Because you could outsmart yourself, don't look for misdirecting questions.

- In any case, picking the most fitting answer may anticipate that you should pay special mind to those answers that are simply mostly right.

If the request is scrutinized circumspectly, beginning presentations are overall right. As, a general rule, don't change an answer with the exception of in the event that you find later that it was misread or aside from if a future request "triggers" your memory.

In the wake of venturing through the tests in this book, check your results. Don't just survey those requests mistakenly answered, yet review each one

The Accreditation Test and This Arrangement Guide

The inquiries for every affirmation test are a firmly protected mystery. Truly on the off chance that you had the inquiries and could just finish the test, you would be in for a significant shame when you showed up at your first occupation that necessary these abilities. The fact of the matter is to know the material, not simply to effectively finish the test. We do understand what points you should know to effectively finish this test since they are distributed by Cisco. Coincidently, these are similar themes needed for you to be capable while designing Cisco IOS switches. It is likewise

Introduction vital to comprehend that this book is a "static" reference, while the course destinations are dynamic. Cisco can and changes the subjects covered on affirmation tests regularly. This test guide ought not be your possibly reference while getting ready for the affirmation test. There is an abundance of data accessible at Cisco.com that covers every theme in difficult detail. The objective of this book is to get ready you as well as could be expected for the SECUR test. A portion of this is finished by breaking a 500-page (normal) usage direct into a 20-page section that is simpler to process. In the event that you imagine that you need more point by point data on a particular subject, don't hesitate to surf. We have separated these themes into establishment points and covered every subject all through the book. Table I-1 records every establishment point alongside a short depiction.

Note that since security weaknesses and protection estimates proceed apace, Cisco Situation maintains whatever authority is needed to change the test goals without notice. Despite the fact that you may allude to the rundown of test targets recorded in Table I-1, consistently keep an eye on the Cisco Frameworks site to check the real rundown of destinations to be certain you are set up prior to taking a test. You can see the current test goals on any current Cisco confirmation test by visiting their site at Cisco.com, clicking Learning and Events>Career Affirmations and Ways. Note likewise that, if necessary, Cisco Press may post extra preliminary substance on the page related with this book at www.ciscopress.com/1587200724. It's a smart thought to check the half a month prior to taking your test to be certain that you have state-of-the-art content.

Table I-1 *SECUR Foundation Topics and Descriptions*

Reference Number	**Exam Topic**	**Description**
1	Secure Administrative Access for Cisco Routers	To ensure that your network is not compromised, it is important to ensure that administrative access to your devices is properly secured. There are several ways to ensure that administrative access to

		Cisco IOS routers is limited to only authorized administrators. The topic is discussed in Chapters 4, 5, and 11.
2	Describe the Components of a Basic AAA Implementation	A successful AAA implementation requires many components. The implementation of AAA is dis- cussed in Chapters 7 and 8.
Reference Number	**Exam Topic**	**Description**
3	Test the Perimeter Router AAA Implementation Using Applicable **debug** Commands	AAA implementation and troubleshooting are ex- plained in Chapters 7 and 8.
4	Describe the Features and	The Cisco Secure Access Control Server is discussed

	Architecture of CSACS 3.0 for Windows	in Chapters 9 and 10.
5	Configure the Perimeter Router to Enable AAA Processes to Use a TACACS Remote Service	The implementation of AAA protocols (TACACS+ and RADIUS) are described in Chapters 7 and 8.
6	Disable Unused Router Services and Interfaces	The most effective way to secure the Cisco IOS router is to disable services and interfaces that are not necessary for the operation of the router. The correct steps for disabling the administrative interfaces are covered in Chapter 5. Disabling unnecessary services is discussed in Chapter 11.
7	Use Access Lists to Mitigate Common	Access lists are a relatively simple way to filter

	Router Security Threats	malicious traffic. The different access list types and configuration steps for each are discussed in Chapter 12.
8	Define the Cisco IOS Firewall and CBAC	CBAC is the basis of the Cisco IOS firewall. Chap- ters 13 and 14 discuss CBAC in great detail and out- line the features of the IOS firewall feature set.
9	Configure CBAC	The configuration of CBAC is explained in Chapter 14.
10	Describe How Authentication Proxy Technology Works	Authentication proxy is a service that enables admin- istrators to proxy user authentication at the firewall. This IOS firewall feature is covered in Chapter 15.
11	Configure AAA on a Cisco IOS	There are many different aspects that all involve

	Firewall	AAA. The configuration of AAA is discussed in Chapters 7, 8, and 9.
12	Name the Two Types of Signature Implementations Used by the Cisco IOS Firewall IDS	The Cisco IDS features on the Cisco IOS firewall are referenced in Chapter 16.
13	Initialize a Cisco IOS Firewall IDS Router	Configuration of the Cisco IOS router IDS is discussed in Chapter 16.
Reference Number	**Exam Topic**	**Description**
14	Configure a Cisco Router for IPSec Using Preshared Keys	VPNs using IPSec and Cisco IOS firewalls are discussed in Chapter 17.
15	Verify the IKE and IPSec	The steps required to verify the configuration of IKE

	Configuration	and IPSec are referenced in Chapter 17.
16	Explain the issues Regarding Configuring IPSec Manually and Using RSA-Encrypted Nonces	The implementation of IPSec using RSA-encrypted nonces is discussed in Chapter 17.
17	Advanced IPSec VPNs Using Cisco Routers and CAs	Configuring VPNs using a certificate authority for peer authentication is a very scalable method for building multiple VPNs. This type of configuration is discussed in Chapter 18.
18	Describe the Easy VPN Server	The Easy VPN Server is defined in Chapter 19. The configuration steps for building VPNs using Easy VPN Server are also covered in this chapter.

19	Managing Enterprise VPN Routers	The products used to centrally manage an enterprise- level VPN using Cisco VPN routers are discussed in Chapter 20.

- Overview of the Cisco Certification Process

- The organization security market is right now in a position where the interest for qualified architects immensely outperforms the inventory. Therefore, numerous designers consider relocating from directing/organizing over to arrange security. Recollect that "network security" is simply "security" applied to "networks." This seems like an undeniable idea; however it is really a vital one on the off chance that you are seeking after your security affirmation. You should be exceptionally acquainted with systems administration before you can start to apply the security ideas. Albeit a past Cisco confirmation isn't needed to start the Cisco security affirmation measure, it is a smart thought to in any event finish the CCNA certificate. The abilities needed to finish the CCNA will give you a strong establishment that you can venture into the organization security field.

- The security confirmation is called Cisco Guaranteed Security Proficient (CCSP) and comprises of the accompanying tests:
- CSVPN – Cisco Secure Virtual Private Organizations (642-511)
- CSPFA – Cisco Secure PIX Firewall Progressed (642-521)
- SECUR – Getting Cisco IOS Organizations (642-501)
- CSIDS – Cisco Secure Interruption Recognition Framework (642-531)
- CSI – Cisco SAFE Execution (642-541)
- The necessities for and clarification of the CCSP affirmation are illustrated at the Cisco Frameworks site. Go to Cisco.com, click Learning and Events>Career Affirmations and Ways.

- **Taking the SECUR Certification Exam**
- **Rules of the Road**

- We have consistently thought that it was befuddling when various locations are utilized in the models all through a specialized distribution. Therefore we will utilize the location space portrayed in Figure I-2 when doling out organization sections in this book. Note that the location space we have chosen is completely held space per RFC 1918. We comprehend that these locations are not routable across the Web and are not typically utilized on external interfaces. Indeed, even with the large numbers of IP tends to accessible on the Web; there is a slight possibility that we might have decided to utilize a location that the proprietor didn't need distributed in this book.

Figure I-2 *Addressing for Examples*

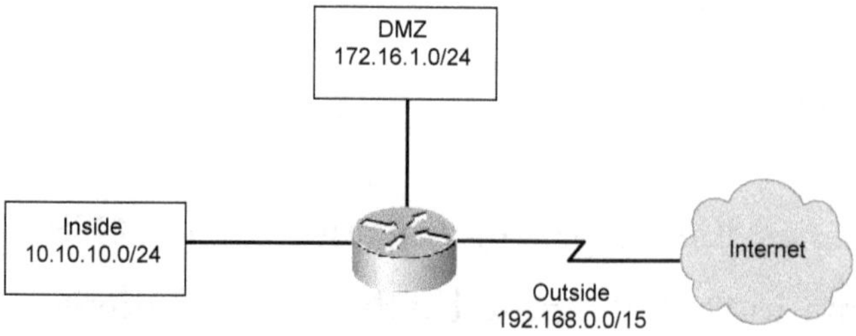

(or any public space)

It is our hope that this will assist you in understanding the examples and the syntax of the many commands required to configure and administer Cisco IOS routers.

Watchwords:

Appurtenance—An advantage or improvement having a spot with, and passing with, the land, For example, an easement appurtenance gives an advantage to one owner to use the other owner's property for passage and flight. Another model would be a home that has a bound garage. For assurance purposes, the separated garage would be seen as an appurtenant plan.

Store of rights—A belonging thought portraying every single one of those authentic rights that join to the obligation regarding property, including the alternative to sell, lease, burden, use, appreciate, preclude, will, etc business specialist—Someone who records and sells associations without the land. They work under rules spread out in the Uniform Business Code. Article 6 of the Code controls mass trades, the proposal of a business by and large, including trade devices, effects, and item. Resource Another name for singular property; resource fuses objects not con-prepared in the significance of real property. Singular property is moved by utilization of a bill of offer. Defeasible cost endowment—An affirmed space in which the grantee could lose his advantage upon the occasion or non-occasion of a foreordained event. There are two sorts of defeasible inheritances, those known as a condition coming about, where the likelihood of reappearance occurs, and an ensured limitation, where the grantee's ownership subsequently gets done with the opportunity of returned (also called a charge essential positive). The words to the extent that, or while, or during are basic to making this second kind of defeasible estate. Devise—the enrichment of authentic property by will. The supplier is the terminated and the recipient is known as a devisee.

Emblements—creating harvests that are conveyed each year through work and industry, also called annuals or fructose industrials. At accumulate time, corn and soybeans would be examples of emblements, and, aside from if regardless agreed, possession would have a spot with the get-together who planted the collect.

Escheat—The reversal of private property to the public expert in circumstances where the decedent passes on without a will or without any recipients fit for obtaining, or when a property is abandoned. Charge clear—the greatest space one can have in certifiable property. A cost fundamental space is the most un-confined interest and the most complete and all out ownership in land: it is of questionable length, straightforwardly versatile, and inheritable.

Mechanical assembly—An article that was once up close and personal property anyway has been so joined to the land that it has become certified property (e.g., broilers, racks, plumbing). At whatever point set out to be an establishment, by then the article passes with the property regardless of the way that it isn't referred to in the deed.

Intestate—to pass on without a genuine will. Not as much as freehold space—A home held by one who rents or leases property. This plan consolidates a permanent place to stay for a significant long time, an irregular residency, a space unreservedly, and a home at lenience. See Area 18 for a natty coarse explanation of such leasehold estates. Life area— any home in certified or individual property that is confined in term to the presence of its owner or the presence of some other appointed person. Lives homes may be made by will or deed, and determine what happens after the finish of the presence inhabitant's life estate. The property could get back to the main grantor or grantor's recipients (life home in reversal) or it could pass in extra part to an assigned remaining portion man (life estate in extra bit). Littoral rights—A landowner's case to usage of the stream coating their property similarly as the use of its shore zone. Littoral rights models consolidate ownership along huge streams like lakes or oceans. Riparian rights are water advantages of owners of land along moving streams, similar to safe streams and streams.

Probate—the formal lawful proceeding to exhibit or attest the authenticity of a will, The will is acquainted with the probate court, and leasers and interested standard ties are advised to acquaint their cases or with show since why the plans of the will should not be approved by the court.

Property—the rights or interests an individual has in the thing asserted; not, in the Particular sense, the genuine thing These rights consolidate the choice to have, to use, to block, to move, and to preclude, usually called the pile of rights.

Certifiable property—all land and appurtenances to land, including structures, structures, mechanical assemblies, divider, and improvements rose upon or affixed to the same, excepting, in any case, creating yields. Riparian rights—those rights and responsibilities that are unplanned to ownership of land coterminous or abutting on conductors, similar to streams and lakes.

Destinations—the financial and individual tendency for one territory of land over another

Loft—Rights in the land that pass with the development. In an easement by common perspective, the servient loft is resolved to allow the overall condo to use their region for passageway and flight, and moves the same obligation to any new buyer for the property. Essentially, the dominating condo gives his choice to continue when the prevalent loft's bundle is sold.

The Best Way to deal with Read for the Land Test

To transform into a real estate agent, you should complete your state's actual licensure test. A land test is by no means, a straightforward test. As a wellspring of viewpoint point, most of all applicants bomb the California land seller test on their first undertaking. Most various states have tantamount pass rates.

Do whatever it takes not to permit this information to stress you. There is inspiring information: You can pass. With fitting examination materials and the right method, you will be set up to float through your test with no difficulty by any stretch of the imagination. Here, our gathering offers the most flawlessly awesome assessment hacks to help you with setting up your territory test.

The Best Tips to Help You Study for the Land Test

Use a Specialist Land Study Guide

While preparing for a land test, there is no convincing motivation to endeavor to sit around idly. It is immovably endorsed that you acquire induction to the best land prep study control. Land tests test a wide extent of different information, from industry-unequivocal language and complex plans to land law and land cash. There are entire courses that are told on all of these topics. All things considered, there is a monstrous proportion of relevant information open. You could analyze it forever.

This is the explanation it is extraordinary get capable land test study materials. Our examination guides take that gigantic area of information and decrease it down to the stuff that truly matters. We created our property test guides thinking about one target: Guarantee that our understudies breeze through the evaluation. There's nothing more to it. Whether or not you are gleaming new to land or you have been in the business for a serious long time, you can benefit by our examination guides. We offer a broad heap of models, test tests, and techniques to help you pass.

Step through Preparing Tests

One of the keys to completing a land test is to take as numerous model tests as you can. Our assistants involve reams of legitimate land test questions. We by and large guarantee that the requests are suitably revived for the current rules, custom fitted to each state, and that they challenge our understudies so they know accurately what's in store during the genuine test. At the point when you start getting most of the preparation tends to right, you will understand that you are ready for test day.

Offer yourself a chance to Plan – Set Step by step and after quite a while after week Study Targets

As you begin perusing for the land test, it is a brilliant idea to set a fundamental course of action of attack. Ideally, you will not be constrained to pack the whole of the information at last. With a lot of workable step by step and after quite a while after week targets, you can persistently set yourself up for the test. This raises a critical issue: How long does it need to peruse for the land test?

There is no one right reaction to this request. Be that as it may, best test-takers spend some place in the scope of 60 and 90 days getting ready for the test. Clearly, this shouldn't suggest that that you are totally stuck between a rock and a hard place if you are endeavoring to pack at last. In case you are as of now spending all accessible time, you need the most perfect land concentrate direct: pick your state and start now.

Review Test-Taking Methods

Finally, when test day moves close, it is a brilliant idea to overview some fundamental test-taking methods. On the edges, acknowledging how to feasibly take a different choice test can help with having a veritable impact. For example, one of the significant frameworks that we recommend is dealing with the hard requests by clearing out the fitting reactions that you know isn't right. Another strategy is to gotten back to questions that are entrapping you and locate all the basic arrangements down first. Past knowing the information – which is undeniably the primary piece of completing a land test – extraordinary evaluation taking affinities can be valuable.

Get Second Permission to the Best Land Test Study Materials

A considerable number understudies have used our advantageous examination associates and free land tests to complete their test on their first endeavor. We are certain that our structure we work for – so much that we offer a full, no requests presented to genuine guarantee. You will complete the test, or we will offer you an all-out markdown. Is it precise to say that you are set up to begin preparing for your state's test? There could be not any more incredible time than today.

CHAPTER 1

Network Security Essentials

The term network security characterizes a wide scope of complex subjects. To comprehend the individual subjects and how they identify with one another, it is significant for you to initially take a gander at the 10,000 foot view and get a comprehension of the significance of the whole idea. Wonder why you lock the way to your home. The appropriate response is likely that you don't need somebody to stroll in and take your stuff. You can consider network security in much a similar design. Security is applied to your organization to forestall unapproved interruptions and robbery or harm of property. For this situation the "property" is "information." In this data age, information has become a truly significant product with both public and private associations making the security of their resources a high need.

Table 1-1 *"Do I Know This Already?" Foundation Topics Section-to-Question Mapping*

Foundation Topics	Questions Covered in This

Section	**Section**
Definition of Network Security	11
Balancing the Business Need with the Security Requirement	9
Security Policies	1, 2, 3, 5, 6, 7, 10
Network Security as a Process	4
Network Security as a Legal Issue	8

CAUTION The objective of self-evaluation is to check your dominance of the subjects in this section. On the off chance that you don't have the foggiest idea about the response to an inquiry or are just part of the way certain about the appropriate response, you should check this inquiry wrong for reasons for the self-evaluation. Furnishing yourself credit for a response you effectively surmise slants your self-appraisal results and may furnish you with a misguided sensation that all is well and good.

1. Which of the accompanying ought to be remembered for the security strategy?

    a) Capabilities of the firewall

    b) Manufacturer of the firewall

    c) User obligations

    d) Sanctions for disregarding the arrangement

    e) A network chart

    f) Routing conventions utilized

2. Which of the accompanying workers ought to approach a duplicate of the security strategy?

3. Managers

4. Network engineers

5. Human assets

6. Temporary workers

7. All workers

8. Which of coming up next is valid about a security strategy?

    a. The strategy ought to require testing.

b. The strategy ought not to be uncovered to the overall population.

c. Cisco hardware ought to be indicated.

d. The strategy is a business report, not a specialized archive.

e. The strategy ought to be changed like clockwork.

9. Which of coming up next are acts coordinated by "the security wheel"?

   a. Configuring

   b. Securing

   c. Implementation

   d. Testing

   e. Monitoring and reacting

10. Which of coming up next are advantages of a security strategy?

   a. Leads to soundness of the organization

   b. Allows the board to sidestep security endeavors

   c. Allows the specialized group to have a limitless

spending plan

d. Enables clients to know the results of their activities

e. Informs the client of how to break into frameworks

11. What are explanations behind executing a security strategy?

a. Enables the board to pass judgment on the adequacy of security endeavors

b. Enables the specialized group to comprehend their objectives

c. Enables clients to peruse the web unafraid of getting an infection

d. Enables the board to legitimize a bigger specialized group

e. Lessens costs because of organization vacation

7. True or Bogus: The security strategy is a record that is intended to permit the business to take part in certain electronic correspondences?

a. True

b. False

8. Choose the six primary objectives of security strategy:

a. Guides the specialized group in buying gear

b. Guides the specialized group in picking their gear

c. Guides the specialized group in arranging the gear

d. Gains the board endorsement for new staff

e. Defines the utilization of the best-accessible innovation

f. Defines the obligations regarding clients and managers

g. Defines sanctions for abusing the approaches

h. Provides a Cisco-focused way to deal with security

i. Defines reactions and accelerations to perceived dangers

9. What is the deciding component while assessing the business need against the security act?

a. Security is consistently the most significant.

b. The business need supersedes security.

c. You need to factor security with the Chime Security Model.

d. Security isn't significant except if your business is sufficiently large to sue.

e. None of the above mentioned.

10. What IETF RFC administers the Site Security Handbook?

a. RFC 1918

b. RFC 2196

c. RFC 1700

d. RFC 1500

11. True or Bogus: Organization security can be accomplished by having advisors introduce firewalls at your organization edge.

a. True

b. False

- The answers to the "Do I Know This Already?" quiz are found in the appendix. The suggested choices for your next step are as follows:

- **8 or less overall score**, Read the entire chapter. This includes the "Foundation Topics" and "Foundation Summary" sections and the "Q&A" section.

- **9 or 10 overall score,** If you want more review on these topics, skip to the "Foundation Summary" section and then go to the "Q&A" section. Otherwise, move on to the next chapter.

CHAPTER 2
Attack Threats Defined and Detailed

This chapter discusses the potential network vulnerabilities and attacks that pose a threat to the network and provides you with a better understanding of the need for an effective network security policy.

"Do I Know This Already?" Quiz

The purpose of the "Do I Know This Already?" quiz is to help you decide whether you really need to read the entire chapter. If you already intend to read the entire chapter, you do not necessarily need to answer these questions now.

The 10-question quiz, derived from the major sections in the "Foundation Topics" portion of the chapter, helps you determine how to spend your limited study time.

Table 2-1 outlines the major topics discussed in this chapter and the "Do I Know This Already?" quiz questions that correspond to those topics.

Table 2-1 *"Do I Know This Already?" Foundation Topics Section-to-Question Mapping*

Foundation Topics Section	Questions Covered in This Section
Vulnerabilities	1, 5
Threats	10
Intruder Motivation	4, 6
Types of Attacks	2, 3, 7, 8, 9

CAUTION The goal of self-assessment is to gauge your mastery of the topics in this chapter. If you do not know the answer to a question or are only partially sure of the answer, you should mark this question wrong for purposes of the self-assessment. Giving yourself credit for an answer you correctly guess skews your self-assessment results and might provide you with a false sense of security.

1. Your boss insists that it is fine to use his wife's name as his password, despite the fact that your security policy states that this is not a sufficient password. What weaknesses are revealed?

    a. This shows a lack of an effective security policy (policy weakness).

b. This shows a technology weakness.

c. This shows a protocol weakness.

d. This shows a configuration weakness.

e. This shows that your boss is an idiot.

2. You receive a call from a writer for a computer magazine. They are doing a survey of network security practices. What form of attack could this be?

    a. Reconnaissance

    b. Unauthorized access

    c. Data manipulation

    d. Denial of service

    e. None of the above

3. Walking past a programmer's desk, you see that he is using a network analyzer. What category of attack should you watch for?

    a. Reconnaissance

    b. Unauthorized access

    c. Data manipulation

d. Denial of service

e. None of the above

4. Looking at the logs, you notice that your manager has erased some system files from your NT system. What is the most likely motivation for this?

a. Intruding for political purposes

b. Intruding for profit

c. Intruding through lack of knowledge

d. Intruding for fun and pride

e. Intruding for revenge

5. Your new engineer, who has very little experience working in your corporate environment, has added a new VPN concentrator onto the network. You have been too busy with another project to oversee the installation. What weakness do you need to be aware of concerning his implementation of this device?

a. Lack of effective policy

b. Technology weakness

c. Lack of user knowledge

d. Operating system weakness

e. Configuration weakness

6. Statistically, what is the most likely launch site for an attack against your network?

    a. From poor configurations on the firewall

    b. From the Internet over FTP

    c. From the Internet through e-mail

    d. From within your network

    e. None of the above

7. Your accountant claims that all the electronic funds transfers from the previous day were incorrect. What category of attack could this be caused by?

    a. Reconnaissance

    b. Unauthorized access

    c. Denial of service

    d. Data manipulation

    e. None of the above

8. Your logs reveal that someone has attempted to gain access

as the administrator of a server. What category of attack could this be?

a. Reconnaissance

b. Unauthorized access

c. Denial of service

d. Data manipulation

e. None of the above

9. Your firewall and IDS logs indicate that a host on the Internet scanned all of your public address space looking of connections to TCP port 25. What type of attack does this indicate?

a. Reconnaissance attack, vertical scan

b. Reconnaissance attack, block scan

c. Reconnaissance attack, horizontal scan

d. Reconnaissance attack, DNS scan

e. Reconnaissance attack, SMTP scan

10. True or False: A "script kiddie" that is scanning the Internet for "targets of opportunity" represents a structured threat to an organization?

a. True

b. False

The answers to the "Do I Know This Already?" quiz are found in the appendix. The suggested choices for your next step are as follows:

- **8 or less overall score**—Read the entire chapter. This includes the "Foundation Topics" and "Foundation Summary" sections and the "Q&A" section.
- **9 or 10 overall score**—If you want more review on these topics, skip to the "Foundation Summary" section and then go to the "Q&A" section. Otherwise, move on to the next chapter.

CHAPTER 3

Defense in Depth

As technology continues to advance, network perimeters are becoming very difficult to define. This chapter looks at the combination of security devices, policies, and procedures required to secure today's networks.

"Do I Know This Already?" Quiz

The purpose of the "Do I Know This Already?" quiz is to help you decide whether you really need to read the entire chapter. If you already intend to read the entire chapter, you do not necessarily need to answer these questions now.

The eight-question quiz, derived from the major sections in the "Foundation Topics" portion of the chapter, helps you determine how to spend your limited study time.

Table 3-1 outlines the major topics discussed in this chapter and the "Do I Know This Already?" quiz questions that correspond to those topics.

Table 3-1 *"Do I Know This Already?" Foundation Topics Section-to-Question Mapping*

Foundation and Supplemental Topics Section	Questions Covered in This Section
Overview of Defense in Depth	1–8

CAUTION The goal of self-assessment is to gauge your mastery of the topics in this chapter. If you do not know the answer to a question or are only partially surc of the answer, you should mark this question wrong for purposes of the self-assessment. Giving yourself credit for an answer you correctly guess skews your self-assessment results and might provide you with a false sense of security.

1. What is the major concern with having a compromised host on the internal network?

   a. It will make the security administrator look bad.

   b. Data on that host can be copied.

   c. Data on that host can be corrupted.

   d. The host can be used to launch attacks against other hosts on the network.

e. None of the above.

2. What are some advantages in implementing AAA on the network? (Choose all that apply.)

a. It limits access to only authorized users.

b. It allows for single sign-on.

c. It provides encrypted connections for user access.

d. It restricts users to only authorized functions.

e. All of the above.

3. Which devices can be used to segment a network? (Choose all that apply.)

a. Firewalls

b. Routers

c. Switches

d. Address scheme

e. All of the above

4. Where does a host based IDS reside?

a. At the network layer

b. At the data link layer

c. At the presentation layer

d. As an add-on to the system processor

e. None of the above

5. What is the advantage of an anomaly based IDS?

a. They protect against unknown attacks.

b. They protect against known attacks.

c. They can restart a Windows server after a system crash.

d. They stop and restart services when needed.

e. They are very cost effective.

6. How does a signature based IDS determine whether it is under attack?

a. It compares the traffic to previous traffic.

b. It compares traffic to predefined signatures.

c. It correlates logs from numerous devices.

d. All of the above.

e. None of the above.

7. Why is it important to monitor system logs?

a. To determine the state of the network

b. To determine whether your systems are running properly

c. To pick a needle from the haystack

d. To determine whether you are under attack

e. To determine whether you can figure out what they mean

8. What is the advantage of using correlation and trending?

a. Most packages print out graphs that you can use for presentations.

b. They enable you to consolidate log data from multiple sources into a readable format.

c. They enable you to correlate log data from multiple sources to get a better understanding of the situation.

d. They enable you to delete traffic that does not apply to your network.

e. None of the above.

The answers to the "Do I Know This Already?" quiz are found in the appendix. The suggested choices for your next step are as follows:

- **6 or less overall score** – Read the entire chapter. This includes the "Foundation Topics" and "Foundation Summary" sections and the "Q&A" section.
- **7 or 8 overall score** – if you want more review on these topics skip to the "Foundation Summary" section and then go to the "Q&A" section. Otherwise, move on to the next chapter.

CHAPTER 4

Basic Router Management

- The Cisco IOS router and Cisco IOS firewall are actually the same hardware. The difference is a low-cost, advanced firewall feature set that was integrated into Cisco Internet Operating System (Cisco IOS). All the basic functionality of Cisco IOS Software remains on the IOS firewall with additional features added, called the firewall feature set. The Cisco IOS router is commonly referred to as the IOS firewall if any of the firewall feature set components are used. This chapter discusses access to and management of the Cisco IOS firewall.

- **"Do I Know This Already?" Quiz**

The purpose of the "Do I Know This Already?" quiz is to help you decide whether you really need to read the entire chapter. If you already intend to read the entire chapter, you do not necessarily need to answer these questions now.

The 10-question quiz, derived from the major sections in the "Foundation Topics" portion of the chapter, helps you determine how to spend your limited study time.

Table 4-1 outlines the major topics discussed in this chapter and the "Do I Know This Already?" quiz questions that correspond to those topics.

Table 4-1 *"Do I Know This Already?" Foundation Topics Section-to-Question Mapping*

Foundation Topics Section	Questions Covered in This Section
Router Configuration Modes	1, 3, 4, 5–8
Accessing the Cisco Router CLI	9, 10
IOS Firewall Features	2

CAUTION The goal of self-assessment is to gauge your mastery of the topics in this chapter. If you do not know the answer to a question or are only partially sure of the answer, you should mark this question wrong for purposes of the self-assessment. Giving yourself credit for an answer you correctly guess skews your self-assessment results and might provide you with a false sense of security.

1. What router configuration mode do you enter by default

when connecting to a router?

a. Console

b. ROM monitor

c. User EXEC

d. Privileged EXEC

e. None of the above

2. Which IOS firewall feature enables you to inspect traffic at multiple layers of the ISO model?

a. Multilayer inspection

b. Context-based access control

c. Tasteful inspection

d. Extended access control lists

e. Connection-based access control

3. Which configuration mode is considered the path to the global configuration mode?

a. User EXEC

b. Line configuration

c. Interface configuration

d. Sub interface configuration

e. None of the above

4. What configuration mode are you in when you see the following prompt on Router A?

a) User EXEC

b) Global configuration

c) Privileged EXEC

d) Unable to determine because the prompt has been changed

e) None of the above

5. What configuration mode must you be in to configure telnet access?

a. Line configuration

b. Interface configuration

c. Telnet configuration

d. Global configuration

e. Connection configuration

f. None of the above

6. What is the default symbol for the global configuration mode?

a. hostname

b. hostname

c. router

d. hostname

e. hostname

7. What command do you use to exit the privileged EXEC mode?

a) Ctrl-Z

b) disable

c) enable

d) exit

e) end

8. What are you most likely doing in the sub interface configuration mode?

a. Changing the telnet password

b. Binding additional IP addresses to an interface

c. Changing the system password

d. Configuring system monitoring

e. Adding the default gateway

9. What access port would you use when connecting a modem?

a. Console port

b. Telnet port

c. Dialup port

d. Secure Shell

e. Auxiliary port

10. What clear-text protocol is not recommended for managing routers from external network segments?

a. Telnet

b. Secure Shell

c. RSH

d. SNMP

e. SMTP

The answers to the "Do I Know This Already?" quiz are found in the appendix. The suggested choices for your next step are as follows:

- **8 or less overall score**—Read the entire chapter. This includes the "Foundation Topics" and "Foundation Summary" sections and the "Q&A" section.

- **9 or 10 overall score**—If you want more review on these topics, skip to the "Foundation Summary" section and then go to the "Q&A" section. Otherwise, move on to the next chapter.

CHAPTER 5

Secure Router Administration

The Cisco IOS firewall helps secure the trusted network from unauthorized users. The security of the network also involves the security of the Cisco IOS firewall itself. In addition to physical security of the Cisco IOS firewall, it is important to secure administrative accesses to interfaces on the Cisco IOS firewall. This chapter discusses the different methods that are available in securing the administrative access to the Cisco IOS firewall.

"Do I Know This Already?" Quiz

The purpose of the "Do I Know This Already?" quiz is to help you decide whether you really need to read the entire chapter. If you already intend to read the entire chapter, you do not necessarily need to answer these questions now.

The 10-question quiz, derived from the major sections in "Foundation Topics" section of the chapter, helps you determine how to spend your limited study time.

Table 5-1 outlines the major topics discussed in this chapter and the "Do I Know This Already?" quiz questions that correspond to those topics.

Table 5-1 *"Do I Know This Already?" Foundation Topics Section-to-Question Mapping*

Foundation Topics Section	Questions Covered in This Section
Secure Administrative Access for Cisco Routers	1–10

CAUTION The goal of self-assessment is to gauge your mastery of the topics in this chapter. If you do not know the answer to a question or are only partially sure of the answer, you should mark this question wrong for purposes of the self-assessment. Giving yourself credit for an answer you correctly guess skews your self-assessment results and might provide you with a false sense of security.

1. What are some of the steps that can be taken to secure the console interface on a router or switch device?

   a. Administratively shut down the console interface.

   b. Physically secure the device.

c. Apply an access list using the **access-class** command.

d. Configure a console password.

2. How many characters can you have in an enable password?

a. 256

b. 32

c. 25

d. 12

3. Which of the following is the least restrictive privilege level?

a. 0

b. 22

c. 15

d. 17

4. The **service password-encryption** command does which of the following?

a. Encrypts the configuration on the router

b. Stores passwords in an encrypted manner in the router

configuration

c. Only encrypts the telnet password in the Cisco IOS configuration

d. Is only available on PIX Firewall

5. Which of the following commands are associated with privilege level 0?

a. Disable

b. configure terminal

c. enable

d. logout

6. Which of the following configurations displays a login banner when a router is accessed?

a. Router# **banner exec** *d If you are not an authorized user disconnect immediately message d*

b. Router **banner login** *d If you are not an authorized userdisconnect immediately d*

c. Router **banner exec** *d If you are not an authorized user disconnect immediately d*

d. Router **banner login** *d If you are not an authorized user* *disconnect immediately d*

7. For maintaining confidentiality and integrity in accessing a router,______is recommended over telnet.

   a. SSH

   b. AH

   c. Secure telnet

   d. VPN

8. How do you secure the Ethernet port on a switch? (Select two.)

   a. Disable unused ports.

   b. Configure port security.

   c. Set access list.

   d. Security cannot be configured on the port.

9. In the event of a security violation, what is the default response of the port?

   a. Switches into restrictive mode

   b. Switches into a temporary shutdown mode

   c. Switches into permanent shutdown mode

d. Switches into a temporary restrictive mode

The answers to the "Do I Know This Already?" quiz are found in the appendix. The suggested choices for your next step are as follows:

- ✓ **8 or less overall score**—Read the entire chapter. This includes the "Foundation Topics" and "Foundation Summary" sections and the "Q&A" section.

- ✓ **9 or 10 overall score**—If you want more review on these topics, skip to the "Foundation Summary" section and then go to the "Q&A" section. Otherwise, move on to the next chapter.

CHAPTER 6

Authentication

- The identification and verification of users requesting access to a device or network is one of the core objectives of security. Although several methods of authentication are available, it is essential that one or a combination of authentication be used to secure the device or network. This chapter provides an introduction to the different types of authentication methods thatyou can use for Cisco devices and networks.

- **"Do I Know This Already?" Quiz**

The purpose of the "Do I Know This Already?" quiz is to help you decide whether you really need to read the entire chapter. If you already intend to read the entire chapter, you do not necessarily need to answer these questions now.

The eight-question quiz, derived from the major sections in "Foundation Topics" section of the chapter, helps you determine how to spend your limited study time.

Table 6-1 outlines the major topics discussed in this chapter and the "Do I Know This Already?" quiz questions that correspond to those topics.

Table 6-1 *"Do I Know This Already?" Foundation Topics Section-to-Question Mapping*

Foundation Topics Section	Questions Covered in This Section
TACACS	5
RADIUS	7
CHAP and PAP	6, 8
Configuring Line Authentication	4
Authentication	1, 2, 3

CAUTION The goal of self-assessment is to gauge your mastery of the topics in this chapter. If you do not know the answer to a question or are only partially sure of the answer, you should mark this question wrong for purposes of the self-assessment. Giving yourself credit for an answer you correctly guess skews your self-assessment results and might provide you with a false sense of security.

1. Which of the following is true? (Choose two.)

   a. Authentication provides a method for verifying the identity of users.

   b. NAS cannot provide authentication.

   c. Usernames and passwords can be stored on NAS.

   d. Cisco does not support RADIUS.

2. Which of the following is the least secure method of authentication? (Choose two.)

   a. Username/password static

   b. Username/password aging

   c. Session key one-time password

   d. Token cards

3. Which of the following security protocols is not supported by Cisco network devices?

   a. TACACS+

   b. RADIUS

   c. Kerberos

   d. TLS

4. Which of the following command syntax is correct for creating a username and password locally on the NAS?

a. Router **username** maroon **password** k0nj0

b. Router **username** maroon **password** k0nj0

c. Router **set username** maroon **set password** k0nj0

d. Router **set username** maroon **password** k0nj0

5. Which port is reserved for TACACS+?

a. UDP 1645

b. TCP 1645

c. TCP 49

d. UDP 49

6. Password Authentication Protocol (PAP)_____

a. Involves a two-way handshake where the username and password are sent across the link in clear text.

b. Sends username and passwords in encrypted format.

c. Involves a one-way handshake.

d. Is not supported by Cisco network devices.

7. Which of the following port does RADIUS use?

   a. UDP 49

   b. TCP 1645

   c. TCP 49

   d. UDP 1645

8. The CHAP authentication protocol ______

   a. Involves a three-way handshake.

   b. Involves a one-way handshake.

   c. Is not supported by Cisco network devices.

   d. Sends password in clear text.

The answers to the "Do I Know This Already?" quiz are found in the appendix. The suggested choices for your next step are as follows:

- **6 or less overall score**—Read the entire chapter. This includes the "Foundation Topics" and "Foundation Summary" sections and the "Q&A" section.

- **7 or 8 overall score**—If you want more review on these topics, skip to the "Foundation Summary" section and then go to the "Q&A" section. Otherwise, move on to the next chapter.

CHAPTER 7

Authentication, Authorization, and Accounting

- An access control system has to be in place to manage and control access to network services and resources. Authentication, authorization, and accounting (AAA) network security services provide the primary framework through which you set up access control on your router or network access server (NAS).

- **"Do I Know This Already?" Quiz**

The purpose of the "Do I Know This Already?" quiz is to help you decide whether you really need to read the entire chapter. If you already intend to read the entire chapter, you do not necessarily need to answer these questions now.

The 10-question quiz, derived from the major sections in "Foundation Topics" section of the chapter, helps you determine how to spend your limited study time.

Table 7-1 outlines the major topics discussed in this chapter and the "Do I Know This Already?" quiz questions that correspond to those topics.

Table 7-1 *"Do I Know This Already?" Foundation Topics Section-to-Question Mapping*

Foundation Topics Section	**Questions Covered in This Section**
Configure AAA on Cisco IOS Firewall	1–6, 9, 10
Test the Perimeter Router AAA Implementation Using Applicable debug Commands	7, 8

CAUTION The goal of self-assessment is to gauge your mastery of the topics in this chapter. If you do not know the answer to a question or are only partially sure of the answer, you should mark this question wrong for purposes of the self-assessment. Giving yourself credit for an answer you correctly guess skews your self-assessment results and might provide you with a false sense of security.

1. Which of the following best describes AAA authentication?

a. Authentication is last defense against hackers.

b. Authentication can only work with firewalls.

c. Authentication is the way a user is identified prior to being allowed into the network.

d. Authentication is a way to manage what a user can do on a network.

e. Authentication is way to track what a user does once logged in.

2. Which of the following best describes AAA authorization?

a. Authorization cannot work without accounting.

b. Authorization provides the means of tracking and recording user activity on the network.

c. Authorization is the way a user is identified.

d. Authorization determines which resources the user is permitted to access and what opera- tion the user is permitted to perform.

3. Which of the following best describes AAA accounting?

a. Accounting is the way that users are identified before they log in to the network.

b. Accounting enables you to track the services users are accessing as well as the amount of network resources they are consuming.

c. Accounting cannot be used for billing.

d. Accounting is a way to curtail where users can go on a network access server.

e. AAA accounting is used only to track users logging on to the network.

4. Which of the following is the correct syntax to specify RADIUS as the default method for a user authentication during login?

a. Authentication radius login

b. login radius authentication

c. aaa login authentication group radius

d. aaa authentication login default group radius

e. radius authentication login

5. Which of the following authorization methods does AAA not support?

a. TACACS+

b. RADIUS

c. SQL

d. NDS

e. Cisco

6. What command enables you to troubleshoot and debug authentication problems?

a. Debug authentication

b. debug AAA authentication

c. authentication debug AAA

d. show authentication

e. show AAA authentication

7. How do you track user activity on your network access server?

a. You cannot track user activities on your NAS.

b. Use AAA authorization only.

c. Use AAA authentication only.

d. A and B.

e. Configure AAA accounting.

8. Which of the following commands requires authentication for dialup users via async or ISDN connections?

a. Authentication default radius

b. authentication default local

c. authentication line isdn

d. authentication login remote

e. authentication radius

9. After an authentication method has been defined, what is the next step to make AAA authentication work on the access server?

a. Set up AAA accounting.

b. Do nothing.

c. Apply the authentication method to the desired interface.

d. Reload the router or NAS.

The answers to the "Do I Know This Already?" quiz are found in the appendix. The suggested choices for your next step are as follows:

- **8 or less overall score**—Read the entire chapter. This includes the "Foundation Topics" and "Foundation Summary" sections and the "Q&A" section.

- **9 or 10 overall score**—if you want more review on these topics, skip to the "Foundation Summary" section and then go to the "Q&A" section. Otherwise, move on to the next chapter.

CHAPTER 8

Configuring RADIUS and TACACS+ on Cisco IOS Software

☐TACACS+ and RADIUS provide a way to centrally validate users attempting to gain access to a router or access server. This chapter discusses the basic configuration of a network access server (NAS) and router to work with TACACS+ and RADIUS servers.

☐"Do I Know This Already?" Quiz

☐The purpose of the "Do I Know This Already?" quiz is to help you decide whether you really need to read the entire chapter. If you already intend to read the entire chapter, you do not necessarily need to answer these questions now.

☐The eight-question quiz, derived from the major sections in the "Foundation Topics" portion of the chapter, helps you determine how to spend your limited study time.

☐Table 8-1 outlines the major topics discussed in this chapter and the "Do I Know This Already?" quiz questions that correspond to those topics.

Table 8-1 *"Do I Know This Already?" Foundation Topics Section-to-Question Mapping*

Foundation Topics Section	Questions Covered in This Section
Configure the Network Access Server to Enable AAA Processes to Use a TACACS Remote Service	1–8

CAUTION The goal of self-assessment is to gauge your mastery of the topics in this chapter. If you do not know the answer to a question or are only partially sure of the answer, you should mark this question wrong for purposes of the self-assessment. Giving yourself credit for an answer you correctly guess skews your self-assessment results and might provide you with a false sense of security.

1. Which of the following is the command to specify the TACACS+ server on the access server?

   a. Takas -server host

   b. takas host

   c. server takas+

d. server host

2. Which is the default port that is reserved for TACACS?

   a. UDP 49

   b. TCP 49

   c. UDP 1046

   d. TCP 1046

3. Which of the following commands enables you to verify or troubleshoot a RADIUS configuration on a network access server?

   a. Show radius

   b. debug radius

   c. debug radius-server

   d. verify radius

4. What is the significance of the **takas-server key** command?

   a. It specifies an encryption key that will be used to encrypt all exchanges between the access server and the TACACS+ server.

   b. It is used to specify a special text when the user logs in to the access server.

c. It is an optional configuration and not required in the TACACS+ configuration.

d. It uniquely identifies the TACACS+ server.

5. Which of the following commands identifies a RADIUS server in a RADIUS configuration?

a. Radius-server host

b. radius-host

c. server radius+

d. server host

6. Which of the following are the basic steps that are required to configure RADIUS on Cisco IOS Software?

a. Enable AAA.

b. Create an access list.

c. Identify RADIUS server.

d. Define the method list using AAA authentication.

7. Which of the following commands deletes the RADIUS server with IP address 10.2.100.64 from a router

configuration?

a. del radius-server host 10.2.100.64

b. remove radius-server host 10.2.100.64

c. no radius-server host 10.2.100.64

d. disable radius-server host 10.2.100.64

8. Which of the following is the default port used by RADIUS?

a. TCP 1685

b. UDP 1645

c. TCP 1645

d. UDP 1685

The answers to the "Do I Know This Already?" quiz are found in the appendix. The suggested choices for your next step are as follows:

- **6 or less overall score** – Read the entire chapter. This includes the "Foundation Topics" and "Foundation Summary" sections and the "Q&A" section.
- **7 or 8 overall score** – If you want more review on these topics, skip to the "Foundation Summary" section and

then go to the "Q&A" section. Otherwise, move on to the next chapter.

CHAPTER 9

Cisco Secure Access Control Server

- Cisco Secure Access Control Server (Cisco Secure ACS) provides AAA services for dialup access, dial-out access, wireless, VLAN access, firewalls, VPN concentrators, administrative controls, and more. The list of external databases supported has also continued to grow, and the use of multiple databases, as well as multiple Cisco Secure ACSs, has become more common.

- This chapter describes the features and architectural components of the Cisco Secure ACS.

- **"Do I Know This Already?" Quiz**

- The purpose of the "Do I Know This Already?" quiz is to help you decide whether you really need to read the entire chapter. If you already intend to read the entire chapter, you do not necessarily need to answer these questions now.

- The 10-question quiz, derived from the major sections in "Foundation Topics" section of the chapter, helps you determine how to spend your limited study time.

- Table 9-1 outlines the major topics discussed in this

chapter and the "Do I Know This Already?" quiz questions that correspond to those topics.

Table 9-1 *"Do I Know This Already?" Foundation Topics Section-to-Question Mapping*

Foundation Topics Section	Questions Covered in This Section
Describe the Features and Architecture of Cisco Secure ACS	1–10

CAUTION The goal of self-assessment is to gauge your mastery of the topics in this chapter. If you do not know the answer to a question or are only partially sure of the answer, you should mark this question wrong for purposes of the self-assessment. Giving yourself credit for an answer you correctly guess skews your self-assessment results and might provide you with a false sense of security.

1. Which of the following devices are supported by Cisco Secure ACS?
    a. Cisco PIX firewall
    b. Cisco Network Access Servers (NAS)

c. Cisco 412

d. Cisco 550

2. Which of the following is true about Cisco Secure ACS?

a. Centralizes access control and accounting

b. Centralizes configuration management for routers and switches

c. Is a distributed security application only for firewalls

d. Only supports Cisco products

3. Which of the following user repository systems are supported by Cisco?

a. Windows NT/2000 user database

b. Generic LDAP

c. Novell NetWare Directory Services (NDS)

d. Cipher Tec database

4. Which of the following password protocols is not supported by Cisco Secure ACS?

a. EAP-CHAP

b. EAP-TLS

c. LEAP

d. ERTP

5. Which of the following is a feature of the Cisco Secure ACS authorization feature?

   a. Denying logins based on time of day and day of week

   b. Denying access based on operating system of the client

   c. Permitting access based on packet size

   d. Permitting access based on the type of encryption used

6. Which of the following are the types of accounting logs that can be generated by Cisco Secure ACS?

   a. Administrative accounting

   b. PAP accounting

   c. TACACS+ accounting

   d. RADIUS accounting

7. Which of the following is not part of the main services/modules that are installed for Cisco Secure ACS for Windows?

   a. CS Mon

   b. CS Admin

   c. CS Auth

   d. CSACS

8. What do the CS Mon services do?

   a. Provides logging services for both accounting and system activity

   b. Provides the HTML interface for administration

   c. Provides recording and notification of Cisco Secure ACS performance.

   d. Monitors firewall activities

9. Authentication and authorization function is handled by which service in the Cisco Secure ACS for Windows?

   a. CS Admin

   b. CS Authen

   c. CS Auth

   d. Secure Authen

10. Under which condition(s), using the Cisco Secure user database, are users forced to change their password?

a. After a specified number of days

b. After a specified number of logins

c. The first time a new user logs in

d. Never

The answers to the "Do I Know This Already?" quiz are found in the appendix. The suggested choices for your next step are as follows:

- **8 or less overall score**—Read the entire chapter. This includes the "Foundation Topics" and "Foundation Summary" sections and the "Q&A" section.
- **9 or 10 overall score**—If you want more review on these topics, skip to the "Foundation Summary" section and then go to the "Q&A" section. Otherwise, move on to the next chapter.

CHAPTER 10

Administration of Cisco Secure Access Control Server

- AAA was conceived originally to provide a centralized point of control for user access via dialup services. As user databases grew, more capability was required of the AAA server. Regional, and then global, requirements became common.

- This chapter provides insight into the deployment process and presents a collection of factors that you should consider before deploying Cisco Secure Access Control Server (Cisco Secure ACS).

- "Do I Know This Already?" Quiz

- The purpose of the "Do I Know This Already?" quiz is to help you decide whether you really need to read the entire chapter. If you already intend to read the entire chapter, you do not necessarily need to answer these questions now.

- The five-question quiz, derived from the major sections in "Foundation Topics" section of the chapter, helps you determine how to spend your limited study time.

- Table 10-1 outlines the major topics discussed in this chapter and the "Do I Know This Already?" quiz questions that correspond to those topics.

Table 10-1 *"Do I Know This Already?" Foundation Topics Section-to-Question Mapping*

Foundation Topics Section	Questions Covered in This Section
Basic Deployment Factors for Cisco Secure ACS	1, 2, 5
Installing Cisco Secure ACS for Windows	3, 4

CAUTION The goal of self-assessment is to gauge your mastery of the topics in this chapter. If you do not know the answer to a question or are only partially sure of the answer, you should mark this question wrong for purposes of the self-assessment. Giving yourself credit for an answer you correctly guess skews your self-assessment results and might provide you with a false sense of security.

1. Which of the following points do you have to consider before deploying Cisco Secure ACS?

    a. Dialup topology

    b. Number of users

    c. Remote access policy

    d. Number of Linux servers

2. Which of the following is the minimum CPU requirement for a Cisco Secure ACS server?

    a. At least a Pentium III 550 MHz

    b. At least Pentium II 330 MHZ

    c. Will work on any Pentium platform

    d. Both A and B

3. Which of the following are task buttons that are present on the web administrative interface of Cisco Secure ACS? How would network latency affect the deployment of Cisco Secure ACS?

   a. User Setup

   b. Group Setup

   c. Network Configuration

   d. System Configuration

4. Which of the following are checklist items that come up during the installation of Cisco Secure ACS?

   a. Windows server can successfully ping AAAclients.

   b. End users can successfully connect to AAA clients.

   c. Your version is at least Netscape version 6.02.

   d. You have a T1 connection.

5. What is the minimum browser version that is supported by Cisco ACS version 3.2?

   a. Netscape 6.02 and Microsoft Internet Explorer 6.0

   b. Mosaic 3.0 and Microsoft Internet Explorer 5.5

c. Netscape 7.0 and Microsoft Internet Explorer 6.0

d. Mosaic 3.0 and Netscape 7.02

The answers to the "Do I Know This Already?" quiz are found in the appendix. The suggested choices for your next step are as follows:

- **3 or less overall score**—Read the entire chapter. This includes the "Foundation Topics" and "Foundation Summary" sections and the "Q&A" section.
- **4 or 5 overall score**—If you want more review on these topics, skip to the "Foundation Summary" section and then go to the "Q&A" section. Otherwise, move on to the next chapter.

All Chapter Answers

Chapter 1

1. a, c, d, e
2. e
3. a, d
4. b, d, e
5. a, d
6. a, b, e
7. True
8. b, c, e, f, g, i
9. e
10. b
11. False

Chapter 2:

1. a, d
2. a
3. a
4. c
5. e
6. d
7. d
8. b
9. c
10. False

Chapter 3

1. d
2. a, b
3. e
4. e
5. a
6. b
7. a, b
8. b, c

Chapter 4

1. c
2. b
3. e
4. d
5. a
6. e
7. b
8. b
9. e
10. f

Chapter 5

1. a, c, d
2. c
3. c
4. b
5. a
6. a, c, d
7. b
8. a
9. a, b
10. c

Chapter 6

1. a, c
2. a, b
3. d
4. a
5. c
6. a
7. d
8. a

Chapter 7

1. c
2. d
3. b
4. c
5. d
6. b
7. b
8. e
9. b
10. c

Chapter 8

1. a
2. b
3. b
4. a
5. a
6. a, c, d
7. c
8. b

Chapter 9

1. a
2. a
3. a, b, c
4. d
5. a
6. a, c, d
7. d
8. c
9. c
10. a, b, c

Chapter 10

1. a, b, c
2. a
3. a, b, c, d
4. a, b
5. c

Conclusion

- The best strategy to complete the PSI Land Test; Fathom Your State's Test Content

- Since you will be taken a stab at your state's property laws and rules despite the more expansive subjects found on the public piece of the land test, you ought to comprehend what governs your state anticipates that you should be familiar with. This information can be found in up-and-comer information handbooks for each state, which are open for download on the PSI website (www.PSIexams.com).

- **Survey Course Materials**

- Numerous states have enlightening requirements that arranged delegates and arrangements experts ought to fulfill before getting able to sit for their approving tests. These can join on the web or grounds based land arrangements, cash and monetary issue courses offered by colleges and state-embraced land schools.

- Since these courses are requested by the state, putting aside exertion to review notes and course materials from your examinations - close by information associated with the up-and-comer information handbook - can help you with developing an assessment plan subject to material covered by the allowing tests.

- **Take a Preparation Test**

- Since PSI passes on the test, you may have to take in any event one practice tests that are directed by the PSI Learning Establishment. You can purchase these preparation tests one by one and take them absolutely on the web. After you complete the preparation test, you will get a score report that offers the correct reactions and an examination of your answer choices. Subsequently, you can use these tests to help perceive your characteristics and inadequacies for future examination.

- In resentment of the way that not supported by PSI, you can similarly find destinations that offer free practice questions or tests that can be used to test your general data and work on reacting to questions.

- **Investigate Open Assessment Resources**

- Test-prep resources are open from a couple of providers. A comparative school or land school you used to fulfill informational prerequisites may similarly offer test-prep courses for public and state fragments of your allowing test. You can moreover take a gander at a segment of oneself assessment resources spread out underneath to sort out which one is ideal for you.

- **PSI Getting ready Framework**

- PSI cooperates with untouchable planning providers, as 360 getting ready, to offer totally online land test-prep that is open for different states. You can take this planning exclusively and locate a consistent speed. Turning into an approved real estate agent incorporates something past acknowledging how to assist someone with finding a respectable home and having amazing capacities in arrangements. Getting ready to transform into a real estate professional incorporates homeroom hours and productive fulfillment of a test to test your knowledge, close by an application cycle to finally get the grant. The system that you choose to peruse for the land licensure and the express that you are living in will choose the schedule opening it takes for you to successfully complete your property licensure.

- States have their own necessities for the amount of hours needed for land planning. For example, Florida requires approximately 63 hours of study corridor coursework and around 45 hours of preparing after you get your grant. If you are in California, you will be expected to complete approximately 135 hours clearly work. Various states, similar to Pennsylvania, may even renounce the land coursework far and away if you hold a long term certificate in land or an associated degree. A couple of states require fundamental courses or school level courses to be done.

- **Land School**

- There are two choices to peruse when perusing for land licensure: on the web or homeroom. Length of study time and cost are two factors to consider while picking which elective best meets your necessities. With the regular homeroom course, you may want to complete the coursework in an ordinary of 4 to a half year. The time frame to complete the coursework will be directed by the state in which you live, as different states have a substitute number of hours required for coursework. Thusly, make sure to check warily what your state needs before you select and start the homeroom bundle.

- In the occasion that you choose to does you're inspecting on the web, you ought to pick guarantee online land school? This course could essentially contract your length of study time, especially if you are prodded and adjust quickly.

- **Land Test and Application**

- Each state has its own necessities for the Land licensure test and application measure. You ought to do a smidgen of investigation to find more about the specific costs for the test and the application in your state. A couple of states even license a short lived grant so you can begin working while you hold on for your position grant. Nevertheless, each state anticipates that you should find and work nearby a confirmed, approved Land vendor when you at first beginning filling in as a Real estate professional.

- **Turning into a Real estate agent Layout**

i. Generally talking, with commitment, focus, troublesome work, one should perhaps have the alternative to transform into a real estate professional inside a year. There are various components to consider with respect to the time slot it will take to transform into an approved real estate professional. Turning into an approved real estate professional incorporates something past acknowledging how to assist someone with finding a fair home and having amazing capacities in arrangements. Getting ready to transform into a real estate professional incorporates homeroom hours and successful satisfaction of a test to test your understanding, close by an application collaboration to finally get the license. The strategy that you choose to peruse for the land licensure and the express that you are living in will choose the schedule opening it takes for you to successfully complete your property licensure. In overview, here is what you should do to transform into a real estate professional:

ii. Examination all costs

iii. Pick a strategy for study

iv. Take and completed supported land courses

v. Effectively complete land test

vi. Apply for land license

vii. Locate an approved land mediator to mentor you

viii. Online Assessment for Transforming into a Real estate agent

- Study.com offers all you need to start toward

transforming into a real estate agent. Courses give total consideration and substance is presented through associating with video activities and text works out. You can in like manner misuse online evaluations to watch your turn of events and get quick permission to dominate teachers through a request and answer feature. Sort out how you can start with Land Planning or explore the Land Test Prep course and study guide or take a gander at the Land Undertakings Manual for start.

- States have their own necessities for the amount of hours needed for land planning. For example, Florida needs around 63 hours of homeroom coursework and about 45 hours of tutoring after you get your license. If you are in California, you will be expected to complete around 135 hours clearly work. Various states, similar to Pennsylvania, may even defer the land coursework totally in case you hold a long term accreditation in land or an associated degree. A couple of states require fundamental courses or school level courses to be done.

- **Land School**

- There are two decisions to peruse when perusing for land licensure: on the web or study corridor. Length of study time and cost are two parts to consider while picking which elective best meets your prerequisites. With the standard homeroom course, you may expect to complete the coursework in an ordinary of 4 to a half year. The time frame to complete the coursework will be directed by the state in which you live, as different states have a substitute number of hours required for coursework. Thusly, make sure to check carefully what your state needs before you select and start the homeroom partition.

- On the remote possibility that you choose to do your considering on the web, you ought to pick guarantee online land school. This course could essentially truncate your length of study time, especially if you are prodded and adjust quickly.

- **Land Test and Application**

- Each state has its own necessities for the Land licensure test and application measure. You ought to do a touch of assessment to find more about the specific costs for the test and the application in your state. A couple of states even license a short grant so you can begin working while you keep things under control for your position grant. Regardless, each state anticipates that you should find and work nearby a guaranteed, approved Land expert when you at first beginning filling in as a Real estate professional.

- **Turning into a Real estate professional Blueprint**

- In general, with responsibility, focus, troublesome work, one should potentially have the alternative to transform into a real estate professional inside a year. There are various segments to consider concerning the interval of time it will take to transform into an approved real estate professional.

CPSIA information can be obtained
at www.ICGtesting.com
Printed in the USA
LVHW010048140821
695291LV00003B/297

9 781802 594041